Title Information:	
Date:	
Logbook #:	
Continued from **Logbook#**	
Name:	
Title:	
Address:	
City & State:	
Email address:	
Telephone #:	
Date Logbook Started:	
Date Logbook Ended:	
Signature	

MW01038764

Notes:-

TABLE OF CONTENTS		
DATE	SUBJECTS	PAGE #

TABLE OF CONTENTS		
DATE	SUBJECTS	PAGE #

TABLE OF CONTENTS

DATE	SUBJECTS	PAGE #

TABLE OF CONTENTS		
DATE	SUBJECTS	PAGE #

	Date	Page #: 2
	___/___/___	Book #:

	Date	Page #: 5
_____	___/___/___	Book #:

	Date	Page #: 10
	___/___/___	Book #:

	Date __/__/__	Page #: 14
		Book #:

	Date	Page #: 18
	___/___/___	Book #:

	Date	Page #: 22
	___/___/___	Book #:

	Date	Page #: 24
	___/___/___	Book #:

	Date __/__/__	Page #: 36
		Book #:

	Date ___/___/___	Page #: 40
		Book #:

	Date __/__/__	Page #: 41
		Book #:

	Date ___/___/___	Page #: 42
		Book #:

	Date ___/___/___	Page #: 63
		Book #:

	Date	Page #: 64
	___/___/___	Book #:

	Date	Page #: 68
	___/___/___	Book #:

	Date	Page #: 82
	___/___/___	Book #:

	Date ___/___/___	Page #: 86
		Book #:

	Date ___/___/___	Page #: 92
		Book #:

	Date ___/___/___	Page #: 94
		Book #:

Made in the USA
Las Vegas, NV
01 October 2023

78403058R10057